Trace, Write, Draw & Learn

SCIENCE WORDS PRACTICE PAGES

GRADES K-2

CLAY MAZE.com

EMPOWERING CHILDREN
FOR A SUCCESSFUL FUTURE

ClayMaze.com

Name ________________________________

Air

1 Trace the words and write them in the space below.

2 Complete the sentence with the word **air** and draw a picture about **air**.

The moving ________ lifted the kite above the trees.

3 Write a sentence using the word **air**.

FUN FACT

Air protects us from rocks falling from space. Shooting stars that you see at night are really meteors burning up in the atmosphere.

Name ______________________________

Animal

1 Trace the words and write them in the space below.

ANIMAL animal

2 Complete the sentence with the word **animal** and draw a picture of an **animal**.

The largest ______________ in the world is the blue whale.

3 Write a sentence using the word **animal**.

FUN FACT Peregrine falcons are the fastest animals and can dive at speeds over 240 mph, but cheetahs have the fastest running speed at 75 mph.

Name ______________________________

Ant

1 Trace the words and write them in the space below.

2 Complete the sentence with the word **ant** and draw a picture of an **ant**.

The ________ lives with its colony underground.

3 Write a sentence using the word **ant**.

FUN FACT **The bullet ant can deliver one of the most painful insect stings in the world, which can be felt for an entire day.**

Name ______________________

Beak

1 Trace the words and write them in the space below.

BEAK beak

2 Complete the sentence with the word **beak** and draw a picture of a **beak**.

A bird uses its __________ to eat and catch food.

3 Write a sentence using the word **beak**.

FUN FACT

The bottom of a pelican's beak has a pouch that is used to scoop up and catch fish. Some can hold up to 3 gallons of water.

Name ________________________________

Bee

1. Trace the words and write them in the space below.

2. Complete the sentence with the word **bee** and draw a picture about **bees**.

The _______ drank nectar from the flower.

3. Write a sentence using the word **bee**.

FUN FACT

A queen honeybee can lay 2,000 eggs in one day.

Name ______________________________

Beetle

1 Trace the words and write them in the space below.

BEETLE beetle

2 Complete the sentence with the word **beetle** and draw a picture of a **beetle**.

The ladybug is a type of ______________.

3 Write a sentence using the word **beetle**.

FUN FACT

Fireflies are not flies, but beetles that light up at night. They can glow many different colors including yellows, greens and reds.

Name ______________________________

Bird

1 Trace the words and write them in the space below.

2 Complete the sentence with the word **bird** and draw a picture of a **bird**.

The mother ________ flew to her nest to feed her young.

3 Write a sentence using the word **bird**.

FUN FACT **The ostrich is the heaviest bird and can weigh over 300 pounds.**

Name ______________________________

Boil

1 Trace the words and write them in the space below.

BOIL boil

2 Complete the sentence with the word **boil** and draw a picture of something **boiling**.

Heat the water to make it __________ and turn into steam.

3 Write a sentence using the word **boil**.

FUN FACT **Water will boil at a lower temperature high in the mountains, because there is less pressure.**

Name ____________________

Bones

1 Trace the words and write them in the space below.

2 Complete the sentence with the word **bones** and draw a picture of some **bones**.

The skeleton is made of ____________.

3 Write a sentence using the word **bones**.

FUN FACT **The smallest bone in the human body is called the stapes and it's in the ear.**

Name ______________________________

Cloud

1 Trace the words and write them in the space below.

CLOUD cloud

2 Complete the sentence with the word **cloud** and draw a picture of **clouds**.

Fog is a type of ____________.

3 Write a sentence using the word **cloud**.

FUN FACT **Clouds are made of water, and although they look light and fluffy, they're actually very heavy and can weigh millions of pounds.**

Name ______________________________

Color

1 Trace the words and write them in the space below.

2 Complete the sentence with the word **color** and draw a picture of something **colorful**.

The ______________ of the plant is green.

3 Write a sentence using the word **color**.

FUN FACT

Bees cannot see the color red, but they can see ultraviolet which people can't see.

Name ______________________________

Core

1 Trace the words and write them in the space below.

CORE core

2 Complete the sentence with the word **core** and draw a picture of a **core**.

There are seeds in the apple's __________ .

3 Write a sentence using the word **core**.

FUN FACT

The Earth's core, located at the center of the planet, is made mostly of iron and is about as hot as the surface of the Sun.

Name ________________________________

Desert

1 Trace the words and write them in the space below.

2 Complete the sentence with the word **desert** and draw a picture of a **desert**.

It is very dry in the _______________.

3 Write a sentence using the word **desert**.

FUN FACT

Not all deserts are hot and sandy. Antarctica is home to the coldest place in the world, and it is also a desert.

Name ______________________________

Dew

1 Trace the words and write them in the space below.

DEW dew

2 Complete the sentence with the word **dew** and draw a picture of **dew**.

There was ________ on the grass this morning.

3 Write a sentence using the word **dew**.

FUN FACT **Insects often drink dew as their source of water.**

Name ____________________

Dinosaur

1 Trace the words and write them in the space below.

2 Complete the sentence with the word **dinosaur** and draw a picture of a **dinosaur**.

That ____________________ is huge!

3 Write a sentence using the word **dinosaur**.

FUN FACT The largest known dinosaur, Argentinosaurus, is thought to have grown to over 100 feet long and weighed over 200,000 pounds.

Name ______________________________

Dolphin

1 Trace the words and write them in the space below.

DOLPHIN dolphin

2 Complete the sentence with the word **dolphin** and draw a picture of a **dolphin**.

The ____________________ jumped out of the water.

3 Write a sentence using the word **dolphin**.

FUN FACT **The killer whale, or orca, is actually a type of dolphin.**

Name ______________________________

Ears

1 Trace the words and write them in the space below.

2 Complete the sentence with the word **ears** and draw a picture of **ears**.

The __________ are used for hearing.

3 Write a sentence using the word **ears**.

FUN FACT

Although ear wax seems gross, it is very important and helps protect our ears from dirt and dust, and also helps to prevent infections.

Name ______________________________

Earth

1 Trace the words and write them in the space below.

EARTH Earth

2 Complete the sentence with the word **Earth** and draw a picture about **Earth**.

______________ is the third planet from the Sun.

3 Write a sentence using the word **Earth**.

FUN FACT **Earth is about four and a half billion years old.**

Name ______________________________

Egg

1 Trace the words and write them in the space below.

2 Complete the sentence with the word **egg** and draw a picture of an **egg**.

The baby bird hatched from the ________.

3 Write a sentence using the word **egg**.

FUN FACT **An ostrich egg would be about as much food as 24 chicken eggs.**

Name ____________________

Energy

1 Trace the words and write them in the space below.

ENERGY energy

2 Complete the sentence with the word **energy** and draw a picture about **energy**.

We use electrical ____________ to power our homes.

3 Write a sentence using the word **energy**.

FUN FACT Venus flytraps are carnivorous plants that get energy and nutrients from eating insects.

Name ______________________________

Eyes

1 Trace the words and write them in the space below.

2 Complete the sentence with the word **eyes** and draw a picture of **eyes**.

We use our ____________ to see.

3 Write a sentence using the word **eyes**.

FUN FACT The colossal squid can have eyes bigger then soccer balls.

Name ___________________________

Feather

1 Trace the words and write them in the space below.

FEATHER feather

2 Complete the sentence with the word **feather** and draw a picture of a **feather**.

The ____________________ was very light.

3 Write a sentence using the word **feather**.

FUN FACT **The feathers on an owl's face help it to hear and locate prey by directing sound toward its ears.**

Name ______________________________

Fins

1 Trace the words and write them in the space below.

2 Complete the sentence with the word **fins** and draw a picture of **fins**.

Fish use __________ to swim.

3 Write a sentence using the word **fins**.

FUN FACT **A thresher shark's tail fin can be as long as the rest of its body.**

Name ___________________________________

Fire

1 Trace the words and write them in the space below.

FIRE fire

2 Complete the sentence with the word **fire** and draw a picture of something on **fire**.

The __________ is very hot.

3 Write a sentence using the word **fire**.

FUN FACT There is a coal fire in Australia that has been burning for thousands of years.

Name ____________________________

Fish

1 Trace the words and write them in the space below.

2 Complete the sentence with the word **fish** and draw a picture of a **fish**.

The ____________ swam very fast.

3 Write a sentence using the word **fish**.

FUN FACT The whale shark is the largest fish and can grow to 40 feet long.

Name ______________________________

Flower

1 Trace the words and write them in the space below.

FLOWER flower

2 Complete the sentence with the word **flower** and draw a picture of a **flower**.

The bee is collecting pollen and nectar from that ____________.

3 Write a sentence using the word **flower**.

FUN FACT Sunflowers can actually move to face the sun.

Name ________________________________

Food

1 Trace the words and write them in the space below.

2 Complete the sentence with the word **Food** and draw a picture of your favorite **food**.

__________ is needed to live and grow.

3 Write a sentence using the word **food**.

FUN FACT A commonly used red food coloring, called carmine, is made from insects.

Name ____________________

Forest

1 Trace the words and write them in the space below.

FOREST forest

2 Complete the sentence with the word **forest** and draw a picture of a **forest**.

There are lots of trees in the __________.

3 Write a sentence using the word **forest**.

FUN FACT **Many medicines we use today come from tropical forest plants.**

Name ______________________________

Fossil

1 Trace the words and write them in the space below.

2 Complete the sentence with the word **fossil** and draw a picture about **fossils**.

That ________________ is very old.

3 Write a sentence using the word **fossil**.

FUN FACT

Poop can be fossilized, and scientists actually study it to learn what creatures ate long ago.

Name ______________________________

Freeze

1 Trace the words and write them in the space below.

FREEZE freeze

2 Complete the sentence with the word **Freeze** and draw a picture of something **frozen**.

______________ the water to make ice.

3 Write a sentence using the word **freeze**.

FUN FACT

When water freezes into ice it gets bigger. Most liquids, not containing water, get smaller when they are frozen.

Name ______________________________

Fruit

1 Trace the words and write them in the space below.

2 Complete the sentence with the word **fruit** and draw a picture of your favorite **fruit**.

The ______________ has seeds inside.

3 Write a sentence using the word **fruit**.

FUN FACT **The tomato is actually a fruit and is one of the most popular fruits in the world.**

Name ____________________

Germs

1 Trace the words and write them in the space below.

GERMS germs

2 Complete the sentence with the word **germs** and draw a picture about **germs**.

I can see ____________ with a microscope.

3 Write a sentence using the word **germs**.

FUN FACT

When you cough or sneeze, germs can travel several feet in front of you. So remember to cover your mouth.

Name ____________________

Gills

1 Trace the words and write them in the space below.

2 Complete the sentence with the word **gills** and draw a picture of some **gills**.

Fish use ____________ to breathe.

3 Write a sentence using the word **gills**.

FUN FACT

Walking catfish have special gills that allow them to breathe air. They can wiggle their way across land and move from one body of water to another.

Name ______________________________

Gravity

1 Trace the words and write them in the space below.

GRAVITY gravity

2 Complete the sentence with the word **gravity** and draw a picture about **gravity**.

The pull of ______________________ keeps us on the Earth.

3 Write a sentence using the word **gravity**.

FUN FACT

Because the Moon has so much less gravity than the Earth, you could jump a lot higher there - probably about 10 feet off the ground.

Name ________________________________

Grow

1 Trace the words and write them in the space below.

2 Complete the sentence with the word **grow** and draw a picture of something that **grows.**

When things ______________ they get bigger.

3 Write a sentence using the word **grow**.

FUN FACT **Some bamboo plants can grow almost 3 feet per day.**

Name ______________________________

Hear

1 Trace the words and write them in the space below.

HEAR hear

2 Complete the sentence with the word **hear** and draw a picture of something you can **hear**.

There are tiny bones in the ears that help us __________.

3 Write a sentence using the word **hear**.

FUN FACT

Dogs can hear high-pitched sounds that people cannot, but cats can hear even higher pitched sounds that dogs can't hear.

Name ______________________________

Heart

1 Trace the words and write them in the space below.

2 Complete the sentence with the word **heart** and draw a picture about the **heart**.

Your ____________ pumps blood through your body.

3 Write a sentence using the word **heart**.

FUN FACT

The heart pumps about 2,000 gallons of blood each day.

Name ______________________________

Heat

1 Trace the words and write them in the space below.

HEAT heat

2 Complete the sentence with the word **Heat** and draw a picture of something **hot**.

______________ the ice to turn it into water.

3 Write a sentence using the word **heat**.

FUN FACT **Sunlight heats up dark objects faster than light ones, because dark colored objects absorb more light than light colored ones.**

Name ______________________

Height

1 Trace the words and write them in the space below.

2 Complete the sentence with the word **Height** and draw a picture about measuring **height**.

______________ is the measure of how tall something is.

3 Write a sentence using the word **height**.

FUN FACT

Gravity causes your height to change while your're up during the day. You are shorter before laying down to sleep and taller in the morning when you wake up.

Name ______________________________

Hive

1 Trace the words and write them in the space below.

HIVE hive

2 Complete the sentence with the word **hive** and draw a picture of a **hive**.

Bees live in a _____________.

3 Write a sentence using the word **hive**.

FUN FACT **Honey bees will often fly 2 to 3 miles from their hive to collect nectar and pollen.**

Name ______________________________

Honey

1 Trace the words and write them in the space below.

2 Complete the sentence with the word **honey** and draw a picture of **honey**.

Bees store ______________ in their hive.

3 Write a sentence using the word **honey**.

FUN FACT

Not all honey tastes the same. The flavor of honey depends on the flower that the bee collected its nectar from.

Insect

Name ______________________________

1 Trace the words and write them in the space below.

INSECT insect

2 Complete the sentence with the word **insect** and draw a picture of an **insect**.

The ________________ has six legs.

3 Write a sentence using the word **insect**.

FUN FACT **The longest recorded insect in the world is a stick bug, which measured over two feet long.**

Name ______________________________

Jupiter

1 Trace the words and write them in the space below.

2 Complete the sentence with the word **Jupiter** and draw a picture about **Jupiter**.

__________________ is the largest planet in our solar system.

3 Write a sentence using the word **Jupiter**.

FUN FACT **Jupiter's largest moon is called Ganymede. It is also the largest moon in our solar system, even larger than the planet Mercury.**

Name ______________________________

Lake

1 Trace the words and write them in the space below.

LAKE lake

2 Complete the sentence with the word **lake** and draw a picture of a **lake**.

There are lots of fish swimming in the ____________.

3 Write a sentence using the word **lake**.

FUN FACT **The deepest lake in the world, Lake Baikal, is over 5,000 feet deep.**

Name ______________________

Larva

1 Trace the words and write them in the space below.

2 Complete the sentence with the word **larva** and draw a picture of a **larva**.

That insect began life as a ____________.

3 Write a sentence using the word **larva**.

FUN FACT Some cicadas live underground as larvae for up to 17 years, but come out to live as adult cicadas for only a few weeks.

Name ____________________________________

Leaf

1 Trace the words and write them in the space below.

LEAF leaf

2 Complete the sentence with the word **leaf** and draw a picture of a **leaf**.

The brown __________ fell slowly to the ground.

3 Write a sentence using the word **leaf**.

FUN FACT

Leaves help plants breathe, but they also help them grow by collecting sunlight energy that can be converted into food.

Name ______________________________

Length

1 Trace the words and write them in the space below.

2 Complete the sentence with the word **length** and draw a picture about measuring **length**.

I used a ruler to measure the ________________ of the line.

3 Write a sentence using the word **length**.

FUN FACT **The longest recorded worm in the world, the bootlace worm, was measured at a length of 180 feet.**

Name ______________________

Light

1 Trace the words and write them in the space below.

LIGHT light

2 Complete the sentence with the word **light** and draw a picture of something **bright**.

The Sun gives off lots of ___________.

3 Write a sentence using the word **light**.

FUN FACT Light is faster than sound. This is why we can see lightning before we hear thunder. Light gets to our eyes faster then sound gets to our ears.

Name ______________________________

Lightning

1 Trace the words and write them in the space below.

2 Complete the sentence with the word **lightning** and draw a picture of **lightning**.

The tree was struck by ______________________.

3 Write a sentence using the word **lightning**.

FUN FACT

Lightning bolts can reach temperatures about 5 times hotter than the surface of the Sun.

Name ______________________________

Liquid

1 Trace the words and write them in the space below.

LIQUID liquid

2 Complete the sentence with the word **liquid** and draw a picture of a **liquid**.

When ice is melted it turns into ____________ water.

3 Write a sentence using the word **liquid**.

FUN FACT Mercury is a shiny metal that is liquid at room termperature, but it weighs over 13 times more than an equal amount of water.

Name ______________________________

Magnet

1 Trace the words and write them in the space below.

2 Complete the sentence with the word **magnet** and draw a picture of a **magnet**.

The ________________ has a north and a south pole.

3 Write a sentence using the word **magnet**.

FUN FACT **Many animals use the Earth's magnetic field to help them navigate.**

Name ______________________________

Mammal

1 Trace the words and write them in the space below.

MAMMAL mammal

2 Complete the sentence with the word **mammal** and draw a picture of a **mammal.**

The dog is a ________________.

3 Write a sentence using the word **mammal**.

FUN FACT **Whales and dolphins are not fish. They are mammals and swim to the surface to breathe air.**

Name ________________________

Mars

1 Trace the words and write them in the space below.

2 Complete the sentence with the word **Mars** and draw a picture about **Mars**.

__________ is sometimes called the 'Red Planet'.

3 Write a sentence using the word **Mars**.

FUN FACT Mars has 2 moons called Phobos and Deimos.

Name ______________________________

Measure

1 Trace the words and write them in the space below.

MEASURE measure

2 Complete the sentence with the word **measure** and draw a picture about **measuring**.

Thermometers are used to ________________ temperature.

3 Write a sentence using the word **measure**.

FUN FACT Some units of measure that we use today are based on the size of body parts. For example, the foot was based on the length of a man's foot.

Name ______________________________

Metal

1 Trace the words and write them in the space below.

2 Complete the sentence with the word **metal** and draw a picture of a **metal**.

The ______________ wire conducts electricity.

3 Write a sentence using the word **metal**.

FUN FACT Pure gold is too soft to use by itself in jewelry, so it is mixed with other metals to make it stronger.

Name ________________________________

Moon

1 Trace the words and write them in the space below.

MOON Moon

2 Complete the sentence with the word **Moon** and draw a picture of the **Moon**.

The __________ has lots of craters.

3 Write a sentence using the word **Moon**.

FUN FACT **The same side of the Moon always faces Earth.**

Name ____________________

Mountain

1 Trace the words and write them in the space below.

MOUNTAIN mountain

2 Complete the sentence with the word **mountain** and draw a picture of a **mountain**.

Snow is melting on the ____________________.

3 Write a sentence using the word **mountain**.

FUN FACT

The longest mountain range in the world is the mid-ocean ridge, and it is mostly under water.

Name ______________________________

Muscle

1 Trace the words and write them in the space below.

MUSCLE muscle

2 Complete the sentence with the word **muscle** and draw a picture about **muscles**.

Exercise builds ______________ and helps it to grow stronger.

3 Write a sentence using the word **muscle**.

FUN FACT There are no muscles in the fingers that control movement. Their movement is controlled by muscles in the palms and forearms.

Name ______________________________

Nest

1 Trace the words and write them in the space below.

2 Complete the sentence with the word **nest** and draw a picture of a **nest**.

The bird laid her eggs in the __________.

3 Write a sentence using the word **nest**.

FUN FACT **The largest bird nest recorded was built by eagles. It was 20 feet deep and weighed over 4,000 pounds.**

Name ______________________________

Nose

1 Trace the words and write them in the space below.

NOSE nose

2 Complete the sentence with the word **nose** and draw a picture of a **nose**.

I can smell with my __________.

3 Write a sentence using the word **nose**.

FUN FACT **Dogs have wet noses to help them smell better, but the wetness also helps them stay cool.**

Name ______________________________

Ocean

1 Trace the words and write them in the space below.

OCEAN ocean

2 Complete the sentence with the word **ocean** and draw a picture about the **ocean**.

The ____________ is very deep.

3 Write a sentence using the word **ocean**.

FUN FACT The deepest part of the ocean, called the Challenger Deep, reaches depths over 36,000 feet.

Orbit

Name ______________________

1 Trace the words and write them in the space below.

ORBIT orbit

2 Complete the sentence with the word **orbit** and draw a picture of something **orbiting.**

It takes the Earth 365 days to ____________ around the Sun.

3 Write a sentence using the word **orbit**.

FUN FACT **It takes the planet Neptune about 165 Earth years to orbit the Sun.**

Name ______________________________

Planet

1 Trace the words and write them in the space below.

2 Complete the sentence with the word **planet** and draw a picture of a **planet**.

Earth is the ____________________ we live on.

3 Write a sentence using the word **planet**.

FUN FACT

Our solar system isn't the only star system with planets. In fact, astronomers believe there are more planets than stars in our galaxy.

Name ______________________________

Plant

1 Trace the words and write them in the space below.

PLANT plant

2 Complete the sentence with the word **plant** and draw a picture of a **plant**.

A ____________ uses it's leaves to collect energy from the sun.

3 Write a sentence using the word **plant**.

FUN FACT

There are over 700 different species of carnivorous, or meat-eating, plants including some pitcher plants that can grow large enough to eat rats.

Name ______________________________

Rain

1 Trace the words and write them in the space below.

2 Complete the sentence with the word **Rain** and draw a picture of **rain**.

__________ helps the plants grow.

3 Write a sentence using the word **rain**.

FUN FACT

On average, it takes about 2 minutes for rain to reach the ground from the clouds, but for small raindrops it can take around 7 minutes.

Name ______________________________

Reptile

1 Trace the words and write them in the space below.

REPTILE reptile

2 Complete the sentence with the word **reptile** and draw a picture of a **reptile**.

The lizard is a ________________.

3 Write a sentence using the word **reptile**.

FUN FACT The largest living reptile (by weight) is the saltwater crocodile, which can grow to over 20 feet long and weigh over 4,000 pounds.

Name ______________________________

River

1 Trace the words and write them in the space below.

2 Complete the sentence with the word **river** and draw a picture of a **river**.

The water is moving quickly down the ___________.

3 Write a sentence using the word **river**.

FUN FACT

The longest rivers in the world are the Nile and the Amazon with lengths over 4,000 miles.

Name ___________________________

Robot

1 Trace the words and write them in the space below.

ROBOT robot

2 Complete the sentence with the word **robot** and draw a picture of a **robot**.

A ____________ that looks like a person is an android.

3 Write a sentence using the word **robot**.

FUN FACT

The Mars rover, called Opportunity, is a robotic vehicle that worked on Mars for over 14 years sending us pictures and information about the planet.

Name ___________________________

Rock

1 Trace the words and write them in the space below.

2 Complete the sentence with the word **rock** and draw a picture of a **rock**.

The mountain is made of ____________.

3 Write a sentence using the word **rock**.

FUN FACT The oldest known rocks on Earth are over 4 billion years old.

Name ______________________

Rocket

1 Trace the words and write them in the space below.

ROCKET rocket

2 Complete the sentence with the word **rocket** and draw a picture of a **rocket.**

The ________________ is ready to launch.

3 Write a sentence using the word **rocket**.

FUN FACT **Fully fueled, the Saturn V rocket weighed over 6,000,000 pounds.**

Name ______________________________

Roots

1 Trace the words and write them in the space below.

2 Complete the sentence with the word **roots** and draw a picture of **roots**.

The plant's ______________ absorb water and nutrients.

3 Write a sentence using the word **roots**.

FUN FACT Some roots can grow over 100 feet deep into the ground.

Name ______________________________

Rotate

1 Trace the words and write them in the space below.

ROTATE rotate

2 Complete the sentence with the word **rotate** and draw a picture of something that **rotates**.

Spin the top to make it ______________.

3 Write a sentence using the word **rotate**.

FUN FACT

Uranus is the only planet in our solar system that rotates on its side.

Name ________________________________

Sand

1 Trace the words and write them in the space below.

2 Complete the sentence with the word **Sand** and draw a picture of **sand**.

________________ is used to make glass.

3 Write a sentence using the word **sand**.

FUN FACT

The tallest sandcastle ever made was over 54 feet tall.

Name ______________________________

Saturn

1 Trace the words and write them in the space below.

SATURN Saturn

2 Complete the sentence with the word **Saturn** and draw a picture of **Saturn**.

There are rings around ______________.

3 Write a sentence using the word **Saturn**.

FUN FACT **Saturn is a giant gas planet, made mostly of hydrogen. It's not solid like Earth.**

Name ______________________________

Science

1 Trace the words and write them in the space below.

2 Complete the sentence with the word **Science** and draw a picture about **science**.

______________ helps us understand the world around us.

3 Write a sentence using the word **science**.

FUN FACT **Astronomy is considered the oldest science. People have been studying the stars and moon for thousands of years.**

Name ______________________________

Seed

1 Trace the words and write them in the space below.

SEED seed

2 Complete the sentence with the word **seed** and draw a picture about **seeds**.

Plant the ____________ in the soil.

3 Write a sentence using the word **seed**.

FUN FACT **A coconut is a giant seed, but it is also considered to be a nut and a fruit.**

Name ______________________________

Shark

1 Trace the words and write them in the space below.

2 Complete the sentence with the word **shark** and draw a picture of a **shark**.

That ______________ has lots of sharp teeth.

3 Write a sentence using the word **shark**.

FUN FACT The whale shark is the largest fish and can grow to 40 feet long.

Name ______________________________

Shell

1 Trace the words and write them in the space below.

SHELL shell

2 Complete the sentence with the word **shell** and draw a picture of a **shell**.

An animal used to live in that sea ______________.

3 Write a sentence using the word **shell**.

FUN FACT

Hermit crabs do not develop their own shells. As they grow, they take abandoned shells from molusks, like sea snails, as their own.

Name ______________________________

Sky

1 Trace the words and write them in the space below.

2 Complete the sentence with the word **sky** and draw a picture of the **sky**.

There are dark rain clouds in the ________.

3 Write a sentence using the word **sky**.

FUN FACT Although it seems white, sunlight is made of different colors. The sky is blue, because the atmosphere scatters blue light more than other colors.

Name ______________________________

Snow

1 Trace the words and write them in the space below.

SNOW snow

2 Complete the sentence with the word **Snow** and draw a picture of **snow**.

__________ is falling from the clouds.

3 Write a sentence using the word **snow**.

FUN FACT It can take over 45 minutes for snow to fall from the clouds to the ground.

Name ______________________________

Soil

1 Trace the words and write them in the space below.

2 Complete the sentence with the word **soil** and draw a picture about **soil**.

I planted the flowers in the __________.

3 Write a sentence using the word **soil**.

FUN FACT There can be more living things in a handful of soil than there are people on Earth.

Name ________________________________

Sound

1 Trace the words and write them in the space below.

SOUND sound

2 Complete the sentence with the word **sound** and draw a picture about a **sound**.

That ________________ is very loud.

3 Write a sentence using the word **sound**.

FUN FACT **Blue whales can hear each other's sounds from hundreds of miles away.**

Name ______________________________

Speed

1 Trace the words and write them in the space below.

2 Complete the sentence with the word **Speed** and draw a picture of something **speedy**.

______________ is a measure of how fast something moves.

3 Write a sentence using the word **speed**.

FUN FACT

The Earth travels around the Sun at a speed of around 67,000 miles per hour.

Name ______________________________

Spider

1 Trace the words and write them in the space below.

SPIDER spider

2 Complete the sentence with the word **spider** and draw a picture of a **spider**.

The ________________ has eight legs.

3 Write a sentence using the word **spider**.

FUN FACT

Although spiders can't fly, some jumping spiders can jump distances 50 times their body length. They use their jumping ability to catch their prey.

Name ________________________________

Spin

1 Trace the words and write them in the space below.

2 Complete the sentence with the word **spin** and draw a picture of something that **spins**.

Can you ______________ a ball on your finger?

3 Write a sentence using the word **spin**.

FUN FACT The world record for the longest spinning top is over 27 hours.

Name ______________________________

Squid

1 Trace the words and write them in the space below.

SQUID squid

2 Complete the sentence with the word **squid** and draw a picture of a **squid**.

The ______________ has 8 arms and 2 longer tentacles.

3 Write a sentence using the word **squid**.

FUN FACT A squid's mouth has a parrotlike beak which it uses to bite and cut its food into pieces.

Name ______________________________

Star

1 Trace the words and write them in the space below.

2 Complete the sentence with the word **star** and draw a picture of a **star**.

The Sun is the closest ______________ to the Earth.

3 Write a sentence using the word **star**.

FUN FACT

Scientists have found a burned out star made of diamond. Its name is Lucy and it's about 2,500 miles across.

Name ______________________________

Sun

1 Trace the words and write them in the space below.

SUN Sun

2 Complete the sentence with the word **Sun** and draw a picture of the **Sun**.

The __________ gives us heat and light.

3 Write a sentence using the word **Sun**.

FUN FACT It takes about 8 minutes for the light from the Sun to reach the Earth.

Name ______________________________

Surface

1 Trace the words and write them in the space below.

2 Complete the sentence with the word **surface** and draw a picture of a **surface**.

The ____________________ of sandpaper is rough.

3 Write a sentence using the word **surface**.

FUN FACT

Water striders are insects that can run across the surface of water, thanks to thousands of tiny hairs that make their legs water resistant.

Name ________________________________

Tongue

1 Trace the words and write them in the space below.

TONGUE tongue

2 Complete the sentence with the word **tongue** and draw a picture of a **tongue**.

I tasted the food with my ________________.

3 Write a sentence using the word **tongue**.

FUN FACT **Cats have a rough spiky tongue that they use to clean and brush their fur.**

Name ______________________________

Tooth

1 Trace the words and write them in the space below.

2 Complete the sentence with the word **tooth** and draw a picture of **teeth**.

I found a shark ______________ on the beach.

3 Write a sentence using the word **tooth**.

FUN FACT **Sharks have many rows of teeth and can lose and replace thousands of teeth throughout their lives.**

Name ____________________

Tornado

1 Trace the words and write them in the space below.

TORNADO tornado

2 Complete the sentence with the word **tornado** and draw a picture of a **tornado.**

The air is swirling around in the ____________________.

3 Write a sentence using the word **tornado.**

FUN FACT **Tornado winds can reach speeds of around 300 miles per hour.**

Name ______________________________

Tree

1 Trace the words and write them in the space below.

2 Complete the sentence with the word **tree** and draw a picture of a **tree**.

The ______________ has a trunk, branches and lots of leaves.

3 Write a sentence using the word **tree**.

FUN FACT

The tallest trees are redwoods, with many reaching heights over 300 feet. The tallest of these trees, named Hyperion, was measured at over 379 feet.

Name ______________________________

Volcano

1 Trace the words and write them in the space below.

VOLCANO volcano

2 Complete the sentence with the word **volcano** and draw a picture of a **volcano.**

There is lava flowing from the ______________________.

3 Write a sentence using the word **volcano**.

FUN FACT **The tallest volcano in our solar system is called Olympus Mons and is located on Mars.**

Name ________________________________

Volts

1 Trace the words and write them in the space below.

2 Complete the sentence with the word **volts** and draw a picture about **volts**.

D batteries generate one and a half ____________ of electricity.

3 Write a sentence using the word **volts**.

FUN FACT **An adult electric eel can generate about 600 volts of electricity, which it uses to catch prey and to deter predators.**

Name ______________________________

Water

1 Trace the words and write them in the space below.

WATER water

2 Complete the sentence with the word **water** and draw a picture of **water**.

We need ________________ to live.

3 Write a sentence using the word **water**.

FUN FACT

The water you drink could be the same water a dinosaur drank millions of years ago.

Name ______________________________

Wave

1 Trace the words and write them in the space below.

2 Complete the sentence with the word **wave** and draw a picture of a **wave**.

A large ______________ crashed on the shore.

3 Write a sentence using the word **wave**.

FUN FACT

Most tsunami waves, which can surge to over 100 feet high, are caused by earthquakes at the bottom of the ocean.

Name ______________________

Width

1 Trace the words and write them in the space below.

WIDTH width

2 Complete the sentence with the word **width** and draw a picture about measuring **width**.

Let's measure the ______________ of the rectangle.

3 Write a sentence using the word **width**.

FUN FACT The Amazon is considered to be the widest river, reaching a width of nearly 7 miles in the dry season and over 24 miles in the wet season.

Name ______________________________

Wind

1 Trace the words and write them in the space below.

2 Complete the sentence with the word **wind** and draw a picture of something **windy**.

The ______________ was very strong during the hurricane.

3 Write a sentence using the word **wind**.

FUN FACT The planet Neptune has wind speeds over 1,000 miles per hour.

Name ________________________________

Worm

1 Trace the words and write them in the space below.

WORM worm

2 Complete the sentence with the word **worm** and draw a picture of a **worm**.

The bird ate the ____________.

3 Write a sentence using the word **worm**.

FUN FACT

Worms eat dead and decaying organic matter like plants and animals, and their poop contains lots of nutrients that help plants grow.

Made in United States
North Haven, CT
05 November 2026

21767754R00059